# Dragons In My Garden

Cath Grimshaw

Dragons In My Garden


National Library of Australia Cataloguing-in-Publication entry (pbk)

| | |
|---|---|
| Author: | Grimshaw, Cath author. |
| Title: | Dragons In My Garden / Cath Grimshaw |
| ISBN: | 9780992581503 (paperback) |
| Subjects: | Children's Fiction |
| | Fantasy Fiction |

Published by Cath Grimshaw and Ocean Reeve Publishing
www.growingdesertroses.com
www.oceanreeve.com

Ocean REEVE PUBLISHING

To my 3 beautiful children

who remind me to believe in dreams.

I was playing in the garden,
Where it would be found,
A tiny little nest,
Hidden on the ground.

It was hidden very well,
I almost didn't see,
That in the tiny nest,
There, three eggs would be.

One, two, three little eggs,
White and almost round,
Covered in dry leaves,
Hidden on the ground.

Something small inside,
What ever could it be?
I am so excited
To see what they will be.

An egg begins hatch,
A little head I see,
A tiny baby dragon
Looking up at me.

Another egg begins to hatch,
Oh, what shall I do?
Suddenly I realized,
My dragons numbered two.

Two little dragons,
One blue, the other pink.
So tiny and adorable,
I even got a wink!

As I sat there watching,
The last egg began to crack,
What colour will it be?
Yellow, orange or black?

Slowly, slowly, hatching, hatching,
A colour I could only dream.
It wasn't yellow, orange or black,
But a beautiful emerald green.

Three baby dragons,
Blue, pink and green,
They are the cutest babies
That I have ever seen!

Now, I need to name you,
I think you are a boy,
A name that means strength,
I think I'll name you Troy.

Now, what will we name you?
I think you are a girl.
So a name that suits you,
I think, would be Pearl.

Oh, little pinky,
No need to be shy,
The name I picked
For you, is Skye.

Endive: A member of the daisy family, this plant is
folate, and

Troy, Skye and Pearl,
Two girls and a boy,
Helping with my homework,
Oh! What a joy!

How do I care for them?
And what do they eat?
They prefer to eat fruit,
And veggies are a treat!

They need time outside,
A safe hiding place,
Amongst the bushy pot plants
To safely play and chase.

Back inside again,
They think they are funny.
Hiding in the cupboard
Behind sauces and the honey.

Oh, they are so cheeky,
These dragons of mine.
I will love them forever
Until the end of time.

When you’re in the garden,
Have a look around,
For a dragons’ nest of eggs,
Hidden on the ground.

Made in the USA
Monee, IL
05 November 2026

67948397R00026